Sugars

by Rhoda Nottridge

Carolrhoda Books, Inc./Minneapolis

Words that appear in **bold** are explained in the glossary on page 30.

Illustrations by John Yates.
Cartoons by Maureen Jackson.

Photographs courtesy of: Aspect, p. 10; Cephas, p. 15; Chapel Studios, cover, pp. 13, 25; Bruce Coleman, pp. 7, 12, 22, 28; Mary Evans, p. 21; Jeff Greenberg, pp. 16, 27; Tony Stone, p. 19; Zefa, pp. 4, 5, 8, 9, 23.

This book is available in two editions:
Library binding by Carolrhoda Books, Inc.
Soft cover by First Avenue Editions
241 First Avenue North
Minneapolis, Minnesota 55401

First published in the U.S. in 1993 by Carolrhoda Books, Inc.

Copyright © 1992 Wayland (Publishers) Ltd., Hove, East Sussex. First published 1992 by Wayland (Publishers) Ltd.

Library of Congress Cataloging-in-Publication Data

Nottridge, Rhoda.
    Sugars / by Rhoda Nottridge.
        p.   cm.
    Includes bibliographical references and index.
    Summary: Focuses on sugars in our diet, explaining where they come from and how our bodies process them. Includes recipes and activities.
    ISBN 0-87614-796-1 (lib. bdg.)
    ISBN 0-87614-611-6 (pbk.)
    1. Sugars in human nutrition—Juvenile literature.
[1. Sugar.] I. Title.
QP702.S85N68   1993
641.3'36—dc20   92-21414
                                                                          CIP
                                                                           AC

Printed in Belgium by Casterman S.A.
Bound in the United States of America

1 2 3 4 5 6 98 97 96 95 94 93

Contents

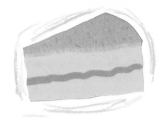

Spotting Sugar

Sugar can make food taste great, and it is a source of quick energy. But eating too much sugar can also be bad for your health. Do you know how many of the foods you eat every day contain sugar?

You probably know that the sweet taste of foods such as cookies, candy, and ice cream comes from sugar. But sugar can also be found in foods that don't taste sweet at all. Check the label of your favorite bread, soup, or spaghetti sauce. You may find sugar listed there.

Sugar is often added to foods for its taste, but sometimes it performs another function. Sugar acts as a **preservative,** which means it prevents foods from spoiling. It is often included in packaged foods and drinks to make them last longer.

It is important to find out which foods contain sugar, because you may be eating more sugar than you realize. Become a sugar spotter. Knowing what foods are made of will help you make healthy choices when you eat.

ABOVE *We eat sugar every day, sometimes without even noticing it. Ice cream tastes great but contains a lot of sugar.*

The sugar family

There are several types of sugar in the sugar family. They all taste sweet, but they don't all come from the same place.

You are probably most familiar with the kind of sugar called **sucrose.** This is the sugar you keep in your sugar bowl at home. Sucrose is also called **refined sugar,** because it is made—or refined —out of sugar beets or sugar-cane. Sucrose contains a mixture of two other kinds of sugar: fructose and glucose.

Fructose is a sugar found in most fruits and many

BELOW
Look at the ingredient label on the side of your favorite canned drink. The word sugar *or* glucose *will probably appear on it. Checking ingredient labels is an easy way to start spotting the sugar in your diet.*

vegetables, and it is the main type of sugar in honey. Fructose is very sweet—about twice as sweet as sucrose. Fruits and vegetables also contain **glucose,** which is about half as sweet as sucrose. Glucose is the kind of sugar that is found in your blood.

Two other kinds of sugar are maltose and lactose. **Maltose** is a sugar formed from grains and other starches. **Lactose** is the sugar found in milk and other dairy products.

What Is Sweetness?

There are about one hundred different things that make foods taste sweet. No one knows why people like sweet tastes but most people do, and so do some animals. Children tend to like sweet tastes even more than adults do.

You taste sweetness with some of your taste buds. A sweet taste is only recognized by the taste buds on the tip of your tongue. Try this experiment.

Stick out your tongue and touch a sugar cube to the side of your tongue. You probably can't taste the sweetness. Wash out your mouth with water. Now place a sugar cube on the tip of your tongue. You will be able to taste the sugar.

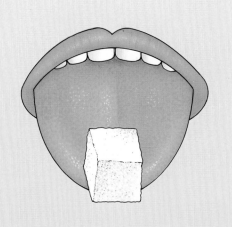

Sugar in plants

You might guess that plants like strawberries and apple trees contain sugar, because their fruits are sweet. But actually all green plants contain sugar, whether they taste sweet or not.

Green plants make sugar by mixing together some of nature's most important materials: light, air, and water. A plant's leaves contain **chlorophyll,** which gives them their green color.

Chlorophyll enables the leaves to absorb light. While the leaves are absorbing light, they are also taking in a part of the air called carbon dioxide. The plant's roots absorb water from the soil. The plant uses the energy in the light to combine the water and carbon dioxide to make the sugar glucose.

Plants need sugar to live and grow. They store it and then use it when they need energy. It also helps them to reproduce. The sweetness of the sugar in fruit attracts birds and other animals. These animals eat the fruit, and the seeds of the plant pass through their stomachs and come out in the animals' droppings. The seeds are scattered around the country-side, and new plants begin to grow.

Most plants don't make very much sugar, so it is only noticeable in their fruit. There are two plants that make and store a lot of sugar. These plants are sugar beets and sugarcane.

BELOW
This fruit bat is dining on wild figs. Unlike humans, animals get all the sugar they need in its natural form from plants and fruit.

Spot That Sugar!

1. Carefully read the ingredient label on your favorite canned or packaged foods and your favorite drink.

2. See if sugar is listed.

3. If the ingredients don't include sugar, see if the words *sucrose, fructose, glucose, maltose,* or *lactose* are listed. These mean sugar too!

4. Make a list of the foods and drinks you've looked at. Note whether they taste sweet and if they contain sugar.

Food	Sweet?	Does it contain sugar?
Hot Dogs	No	Sugar
Orange Soda Pop	Yes	Glucose

Growing Sugar

Sugarcane

Sugarcane is a very tall grass that grows in hot, sunny areas where there is plenty of rainfall. The canes have long, golden stalks that grow up to 15 feet high.

The plant stores its sugar supply inside the stalk. In some places, you can buy chunks of sugarcane to eat. You can taste the sugar as you suck on the cane.

Sugarcane can be harvested at any time when the ground is dry and the farmer has decided the canes are big enough to contain a large store of sugar.

The canes used to be cut down by hand, one by one. This was hot, tiring work. Nowadays machines are usually used to do the hard work of cutting down the canes.

The roots of the plants are left in the ground after the sugarcane has been harvested. The next year, new plants grow from the old roots.

RIGHT *The sugarcane plant stores its sugar in the cane. At this stage, the sugar is very different from the sugar we use every day.*

Sugar beets

Sugar beets are root vegetables, like potatoes and carrots. They store large amounts of sugar in a white root that is the shape of a thick carrot.

Sugar beet plants are grown from seeds. These seeds come from plants that have been left in the ground until their second year so they will flower and produce seeds. The plants that are used for sugar are harvested after their first year.

Unlike sugarcane, sugar beets grow well in places where it is warm in the summer and cool or cold in the winter. Just under half of the world's refined sugar comes from sugar beets. The rest comes from sugarcane.

BELOW
When the sugar beet is harvested, the leafy green tops that grow above the soil are cut off and used as animal feed.

Investigation

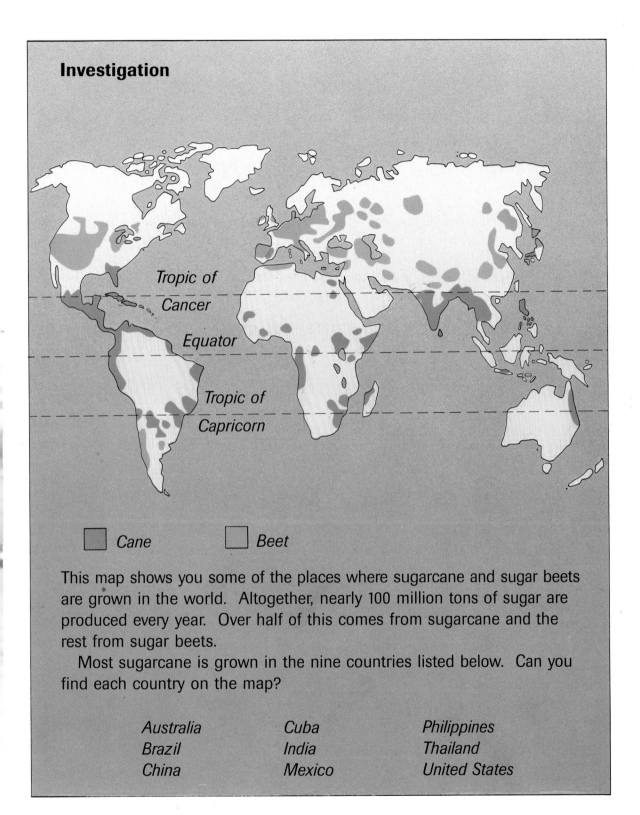

Tropic of
Cancer

Equator

Tropic of
Capricorn

☐ Cane   ☐ Beet

This map shows you some of the places where sugarcane and sugar beets are grown in the world. Altogether, nearly 100 million tons of sugar are produced every year. Over half of this comes from sugarcane and the rest from sugar beets.

Most sugarcane is grown in the nine countries listed below. Can you find each country on the map?

Australia	Cuba	Philippines
Brazil	India	Thailand
China	Mexico	United States

The Sugar Factory

ABOVE
Modern sugar mills, such as this one in South Africa, process thousands of tons of sugarcane or sugar beets every day.

To get all the sugar out of sugar beets and sugarcane, the plants have to be taken from the farms where they grow to a factory. In the factory, they are changed from plant material into the small crystals we know as sugar.

At the factory, the sugar beets or stalks of sugarcane are cut into small pieces.

Then the pieces are placed in huge tanks filled with hot water. The sugary juice from the beets or the cane dissolves in the water, just as sugar dissolves in drinks like tea or coffee.

Another way to get the juice out of sugarcane is by squeezing the canes in crushing machines. The cane is then sprayed with water to remove any remaining juice.

The little bits of squashed sugarcane that are left over are used as fuel or made into a pulp that is used to make paper. The leftover sugar beets can be used as animal feed.

The sugar mixture is muddy and dark, and it needs to be purified. The mixture is heated and then combined with a chemical called lime. The lime separates the sugar from any impurities, leaving behind a pure solution of sugar and water.

The sugar solution is heated in a special boiler. It becomes so hot that most of the water evaporates. What is left is a sugary syrup.

The syrup then has to be turned into sugar crystals. Crystals will form when they have other crystals to attach themselves to. So, sugar crystals are added to the syrup to form new crystals.

The crystals are then separated from the remaining

BELOW
The sugar-making process produces many different types of sugars and syrups. This small selection shows only a few of the sugar products on sale in supermarkets.

liquid by rotating the mixture in a huge drum. A thick syrup called **molasses** is drained off.

Refining sugar

The sugar crystals that come out of the huge spinning drums still have a fine coating of molasses on them. These large, yellowish-brown crystals are called **raw sugar.**

The raw sugar goes to another factory where it is refined, which means that it is made even more pure. The sugar is soaked in a thick, syrupy sugar solution. The sugar solution is **saturated.** This means that the mixture is so sugary already that it cannot dissolve any more sugar or make any more sugar crystals.

The coating of molasses comes off the crystals and sticks to the syrup. Then the crystals and the syrup are spun around together in a machine called a centrifuge to separate them.

The crystals are dissolved in water and are allowed to form into crystals again. This time the crystals are much smaller. This is the pure, white refined sugar that you use every day.

Making a Saturated Solution

Sugar crystals dissolve in water and make a sugary solution. However, there comes a point when so much sugar has dissolved in the water that no more can dissolve in it. No matter how hard you try, you will not be able to dissolve any more crystals. This is called a saturated solution. This is the kind of solution that is used to remove the molasses from sugar crystals when sugar is being refined.

1. Fill a glass half full of warm water.
2. Add a very small pinch of granulated sugar crystals.
3. Stir the solution rapidly until all the crystals have dissolved.
4. Keep stirring in crystals, a few at a time. Eventually, the crystals will not dissolve anymore and will stay at the bottom of the glass. The solution is saturated.

Energy from Sugar

Sugar may be a quick source of energy, but is it a good one?

Some sugars, such as maltose, sucrose, and lactose, need very little digesting. Other sugars, including fructose and glucose, need no digesting at all. So, when you eat sugar, it enters your bloodstream very quickly and you get a burst of energy. But this energy lasts for a very short time. A sudden rise in blood sugar is followed

BELOW
Some people become addicted to eating sugary foods such as chocolate and experience strong cravings for them.

by a sudden drop. Sometimes you end up feeling more tired than you did before you ate the sugar.

Some people find that once they start eating sugar, they want more and more. Their energy levels rise and fall and rise again as they satisfy their cravings for sweets. Not only is this hard on their bodies, but it can also make them feel sick.

Many people are concerned

about the effects of refined sugar on their bodies. When sugar is refined, all the vitamins and minerals are removed. People often eat other sugars, such as raw sugar, instead of refined sugar. But although these sweeteners do have some nutritional value, it is very slight. And, just like refined sugar, too much of them can cause a rapid rise and drop in blood sugar.

Rather than replacing refined sugar with other sugars, it is probably smarter to look to other foods for energy. Foods like fresh fruit, whole grain breads, and pasta are very good for you, and they'll give you all the energy you'll ever need.

ABOVE
Drinking sugar-free beverages is one way of cutting out extra sugar.

Light Sugars

These are all refined sugars.

Granulated white sugar
This is the most commonly used sugar. It is also the cheapest.

Cube sugar
This is made from granulated sugar that has been dampened and then molded into cube shapes.

Superfine sugar
This is like granulated sugar but the crystals are smaller and dissolve more easily. It is often used in baking.

Powdered sugar
This is sugar crystals that are ground into a fine powder. It dissolves quickly in liquids to make a creamy paste that can be used as a frosting.

Canning sugar
This is used to make jam. It has large crystals that dissolve very slowly, which prevents the jam from burning.

Dark Sugars

Brown sugar
This is white sugar that has had molasses added back to it. The darker the sugar, the more molasses it contains. Brown sugar is a refined sugar.

Molasses
This is the dark brown liquid that is separated from the sugar crystals at the factory. Molasses is good for you. It contains calcium, iron, potassium, and B vitamins.

Syrups
Some of the sugar solutions left over from refining sugarcane or sugar beets can be made into golden syrups. There is also another kind of syrup, which comes from the sap of a maple tree. This is called maple syrup. Other sugary syrups are made from vegetables such as corn.

The History of Sugar

The exact history of sugarcane, the first known source of sugar, is not clear. It probably grew wild in the Polynesian Islands, far out in the Pacific Ocean. From there the magical sugarcane was somehow brought to India and China, along with the secret of how to use its sweetness. There came to be stories about sugarcane. In India a king called Subandu is said to have found sugarcane growing in his bedroom. The sugarcane produced a prince named Ilshvaku.

LEFT *In the past, sugar was a luxury for the wealthy. Now many products made from sugar, like these candies, can be bought for just a few cents.*

The secret of sugarcane

The history of sugar is a fascinating tale of a food that has brought fabulous wealth to some and unhappiness and suffering to others.

In 510 B.C., people from Persia, which is now called Iran, invaded India and discovered the tall canes growing there. They were delighted with this wonderful plant that gave "honey without bees." They took sugarcane plants back to Persia. Over hundreds of years, they became the most skilled growers and makers of sugar, jealously guarding the secret of the canes.

About 200 years later, Alexander the Great conquered much of Asia, including Persia. He then founded the city of Alexandria in Egypt. The people of Alexandria began growing sugar and selling it. They used crushed sea snails, insects, and herb roots to color the sugar violet or pink. The ancient Greeks and Romans thought of sugar as a luxury and even used it as a medicine.

The popularity of sugar spread rapidly. Sugarcane was grown in such places as Cyprus, Rhodes, southern Spain, and Syria. By the A.D. 900s, ships loaded with sugar sailed across the Mediterranean Sea to sell the new luxury to merchants in Venice.

During the Crusades, sugar was introduced to Britain. The Crusades were an attempt by Christians to claim the Holy Land in the Middle East. In 1099 British crusaders traveling through Syria came across something they thought was a new spice. They eagerly brought their discovery—sugar—home with them.

Sugar and slavery

By the 1400s, the explorer Christopher Columbus had taken sugarcane to the West Indies. To his delight, he discovered that this was the perfect place to grow the valuable plant, with just the right amount of rain and sun. Europeans began to experiment with growing sugarcane

ABOVE
In the early days of sugar production, people from Africa were taken from their homes and forced to work in terrible conditions as slaves in sugar factories.

Many died at sea, and many more died as slaves working on the plantations.

At that time, there were no machines to make the work easier. The slaves worked in terrible conditions under the burning sun, cutting down the canes by hand. The Europeans grew richer and richer. Plantations continued to be worked by slaves until the 1800s, when slavery was gradually abolished.

in other countries too, such as Brazil, Cuba, and Mexico.

Dutch and French explorers built sugar **plantations** in Indonesia, the Philippines, and Hawaii. Before long, owning a sugar plantation was a way to become enormously wealthy. Sugarcane became known as white gold.

However, these sugar plantations succeeded only because of the efforts of the slaves who were forced to work for the plantation owners. To find people to work on the plantations, the Europeans turned to Africa. People were torn away from their homes and their villages. They were put in chains and forced onto ships traveling to places like the West Indies.

The sugar beet story

Sugar beets were eaten as a vegetable as far back as Roman times but it was not until 1812 that a good method was found of getting the sugar out of the beet plant.

The French and the British had been at war for some time. The British took control of the seas and stopped the French from bringing any sugar into France. Napoleon, the French leader, encouraged the French to grow sugar beets so that they would no longer have to rely on sugarcane.

By 1880 sugar beets had become the main source of sugar in Europe, and it was also grown in many other countries, including the United States and Canada.

Sugar and Health

Sugar can have a long-term effect on your health. There are two very important reasons for not eating too much sugar. The first reason is that sugar can harm your teeth.

BELOW *It is healthiest to eat food containing sugar in its natural form. A crunchy apple makes a tasty snack, and it is much better for you than sweets.*

When you eat sugar, some of it stays in your mouth and becomes part of a sticky coating on your teeth called **plaque.** The plaque slowly builds up between your teeth and around your gums.

There are tiny germs called bacteria living in your mouth. They feed on the particles of food in plaque, especially sugar, and turn it into an acid. This acid softens the layer of enamel that protects your teeth, and so your teeth start to decay.

The best way to avoid tooth decay is by not eating and drinking sugary things. Brushing teeth often will help to get rid of plaque. It is also important to get your teeth checked regularly by a dentist.

The second reason not to eat too many sugary foods is that it can make you overweight. Some people gain a lot of weight because they eat too much sugar and do not exercise. Their bodies have no way of using up all the sugar they eat, so it is stored as fat. This can lead to serious illnesses such as heart disease or even heart failure.

Some people suffer from an illness called **diabetes,** which means their bodies have trouble processing the sugar in their blood. For them, eating sugar can be dangerous. They have to be very careful to avoid sugar in their food.

Cutting down on sugar
It is worth getting into the habit of thinking about how much sugar there is in your food. You do not need to worry if you eat sugary things every now and then, but try to change some of your everyday eating habits. There are many ways of cutting down on the amount of sugar in your diet.

Why not choose unsweetened fruit juices and drink these diluted with water. An ordinary soft drink contains around 5 teaspoons of sugar!

Checking Your Teeth for Plaque

Normally, you can't easily see plaque because it is colorless. But disclosing tablets, which you can buy at the drugstore, can help you to see it. Chew one of these tablets. It will turn parts of your teeth bright pink. The pink spots are to show you where you have plaque. When you brush off the pink spots, you will be brushing away the plaque. Chew a second disclosing tablet when you have brushed your teeth. Are there any pink spots left? If there are, brush your teeth again to remove the last bits of plaque.

Remember, though, that it is best to cut down on sugary foods and drinks to avoid getting plaque in the first place.

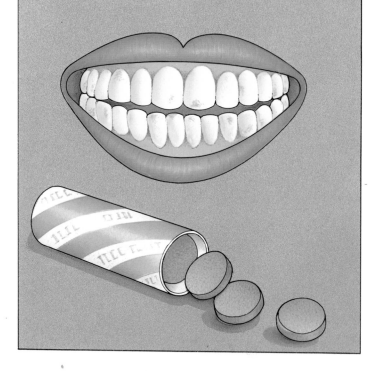

If anyone in your family drinks tea or coffee, suggest that they slowly cut out added sugar. They may even like it better that way. Check that the breakfast cereal you eat does not have added sugar. Some cereals contain so much sugar that they are half sugar and half cereal. Try to find a cereal that does not contain any sugar and see if you can persuade your family to change to it.

Cutting down on sugar also means choosing snacks that contain very little sugar and eating fewer sugary desserts. You can eat plenty of fresh fruit, vegetables, crackers, milk, unsweetened yogurt, or nuts instead.

BELOW *You can sweeten your breakfast cereal naturally with chopped fruit such as bananas and strawberries instead of using refined sugar.*

Sweeteners Today

There are other kinds of sweeteners available besides sugar. Nowadays a lot of people use **artificial sweeteners** such as saccharin. This kind of sweetener is very popular in soft drinks. If you check the label of any food labeled "diet," you will probably find that it contains an artificial sweetener.

Artificial sweeteners are made from chemicals and are very recent inventions. Scientists don't yet know if they have any bad effects on the human body, because the sweeteners haven't been around long enough to test them completely.

Honey is probably the oldest source of sweetness known to humans. It is the sweet, thick liquid made by bees. It contains the sugars fructose and glucose.

RIGHT
For thousands of years, honey has been used as a natural way to sweeten food.

To make honey, bees sip a liquid called **nectar** from flowers. The bees then carry this nectar back to their hives in a pouch in their bodies. The nectar in the pouch breaks down into fructose and glucose.

The bees store the sugar mixture in tiny compartments, or cells, in their hives. There the water in the sugar mixture evaporates. What is left turns into honey.

Honey contains a small amount of the minerals your body needs, including potassium, calcium, and phosphorus. For this reason, many people use honey instead of refined sugars or artificial sweeteners.

ABOVE
This honeybee is filling its pouch with nectar. The bee will return to the hive to store the nectar.

Recipe

Here is a recipe for Yogurt Snow. It uses honey as a sweetener instead of white sugar. If you are concerned about eating raw eggs, you can leave them out.

You will need:
2 egg whites
3 tablespoons honey
2 cups unsweetened yogurt

1. Carefully separate egg whites from egg yolks.
2. Beat egg whites until they are stiff.
3. Slowly add honey and continue to whisk until mixture is stiff.
4. Carefully fold in yogurt.
5. Serve as a topping for fruit salad or a dessert.

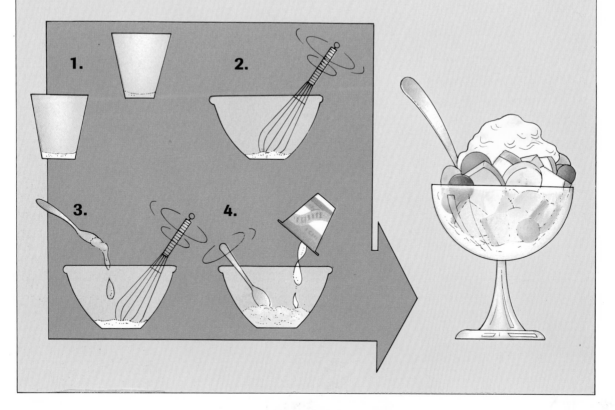

Glossary

Artificial sweeteners Manufactured chemicals used to sweeten food

Bacteria Microscopic organisms that can live in your body

Chlorophyll A substance found in plants that absorbs light and also gives plants their green color

Diabetes A disease in which the body can't process sugar

Fructose A sugar found in all fruits and some vegetables

Glucose A sugar found in fruits and vegetables. It is also the sugar in your blood.

Lactose Milk sugar

Maltose A sugar formed from grains

Molasses The thick, brown syrup that is separated from sugar when sugar is refined

Nectar A sweet liquid produced by plants. Bees use nectar to make honey.

Plantation A large farm, usually devoted to one crop

Plaque A sticky film composed of bacteria, food, and saliva that forms on your teeth

Preservative A substance that prevents a food from spoiling

Raw sugar Large, yellowish-brown crystals of partially purified sugar

Refined sugar Sugar that has been processed so it contains pure sucrose

Sucrose Table sugar. Sucrose contains fructose and glucose.

Books to Read

Food by David Marshall (Garrett Educational Corporation, 1991)

Marceau Bonappétit by Fanny Joly-Berbesson (Carolrhoda Books, 1989)

Sugaring Season: Making Maple Syrup by Diane Burns (Carolrhoda Books, 1990)

Metric Chart

To find measurements that are almost equal

WHEN YOU KNOW:	MULTIPLY BY:	TO FIND:
AREA		
acres	0.41	hectares
WEIGHT		
ounces (oz.)	28.0	grams (g)
pounds (lb.)	0.45	kilograms (kg)
LENGTH		
inches (in.)	2.5	centimeters (cm)
feet (ft.)	30.0	centimeters
VOLUME		
teaspoons (tsp.)	5.0	milliliters (ml)
tablespoons (Tbsp.)	15.0	milliliters
fluid ounces (oz.)	30.0	milliliters
cups (c.)	0.24	liters (l)
quarts (qt.)	0.95	liters
TEMPERATURE		
Fahrenheit (°F)	0.56 (after subtracting 32)	Celsius (°C)

Index

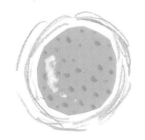